Ecology and Genetics

An Essay on the Nature of Life and the Problem of Genetic Engineering

Ecology and Genetics:

An Essay on the Nature of Life and the Problem of Genetic Engineering

Arjun Makhijani

A report of
The Institute for Energy and
Environmental Research

The Apex Press

Published by The Apex Press, an imprint
of the Council on International and Public
Affairs, Suite 3C, 777 United Nations Plaza,
New York, NY 10017 (800-316-2739)

Library of Congress Control Number: 2001089619

ISBN 1-891843-12-5

Cover design and typesetting by Peggy Hurley

Cover illustrations: circle-shaped DNA by Peggy Hurley, based on diagrams of John W. Kimball; double-helix DNA from: MOLECULAR CELL BIOLOGY by H. Lodish, A. Berk, S. L. Zipursky, P. Matsudaira, D. Baltimore, and J. Darnell © 1986, 1990, 1995, 2000 by W. H. Freeman and Company. Used with permission.

Printed in the United States of America

Table of Contents

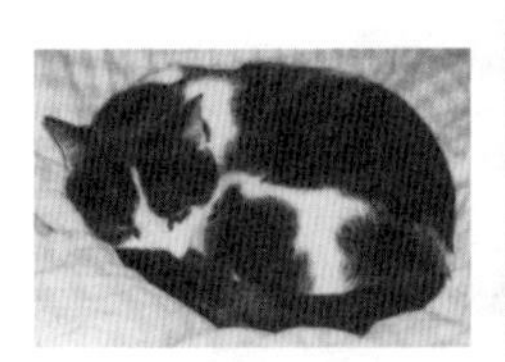

To M-and-m and Caline,
the cats in our family

Preface

I have long dealt with another kind of nuclear pollution—the radioactive mess created by nuclear weapons production and commercial nuclear power. Worrying about radioactive dumps and spent nuclear fuel laced with plutonium-239 has occupied much of my professional life. A large part of my concern has arisen from the longevity and particular danger of plutonium-239. It has a half-life of more than 24,000 years, it is a peril to health, and it can be used to make nuclear weapons.

Plutonium will burden generations far into the future. It is, moreover, a burden whose nature we cannot properly grasp. In the meantime, the risks of nuclear war, by design or accident, continue to hang over humanity. Indeed, the risks of accidental nuclear war have grown.[1]

The implicit social ethic that has produced such risks is the opposite of how people have lived from time immemorial—that society should try to leave a world to the next generation at least as livable as the one it inherited. We live in a world where the pre-occupation with the well-being of the present generation—increasing its comfort and longevity—has become so intense that it is pushing out and severely compromising the possibility for future generations to live well, or even to exist. The immense nuclear arsenals that still exist and the mounting global stocks of plutonium are only a

---

[1] Back from the Brink Campaign and Project on Participatory Democracy 2001.

part of the evidence for such a conclusion.

The problem of inter-species genetic engineering seems quite parallel to that of plutonium. We are proceeding on the basis of slim knowledge and little understanding. We are courting disaster in ways that are ill-understood. Indeed, some theoretical arguments indicate that it may not be possible to understand the consequences of widespread introduction of genetically engineered species without experiments that risk unpredictable destruction.

Plutonium and long-lived nuclear wastes were produced with the full confidence that they could be well managed. Society was supposed to make a "Faustian bargain" with the nuclear establishment, as the former head of the Oak Ridge National Laboratory, Alvin Weinberg, put it in 1972. If society would only trust that the "nuclear people" would manage nuclear waste properly, it would be provided with an endless "magical energy source" from plutonium. The problems of nuclear weapons, of proliferation, of long-term waste management would all be taken care of.[2] The reality has been rather more difficult and complex.

The corporations that are making genetically engineered seeds are making an essentially similar appeal. We are being asked to trust that they will safeguard us from adverse consequences of the broadcast of genetically engineered organisms. We are being asked to trust that the consequences are known well enough and that they will be minimal. In return

---

[2] Weinberg 1972.

for this trust, society will reap an era of plenty in food and freedom from starvation. There is little basis for such trust, if we are to go by the facts of the terrible inequities that have plagued the world for centuries, and which persist to this day.

While long-lived nuclear waste is one of the most vexing and difficult problems that humanity has created, it is still possible, though not without considerable chutzpah, to speak about its management. Almost all (~99 percent) the plutonium that exists is in the custody of various nuclear establishments. Yet the collapse of the Soviet Union has reminded us, were a reminder necessary, that the stability of nuclear establishments is nowhere near the longevity of plutonium. Even the small fraction that has been dumped or that is unaccounted for will remain a concern for health, environment, and security for the foreseeable future.

A similar claim of control cannot be made in regard to the new species created by inter-species genetic engineering, even though it is far more recent than nuclear weapons or nuclear power technology. As biologist Erwin Chargaff has noted, "you cannot recall a new form of life."[3] Further, the potential for inter-species genetic engineering to create new, unmanageable problems of biological warfare and proliferation is just beginning to be glimpsed.

I am emboldened to put forward this essay, though I am not academically trained as a biologist, because those who have seen it, among them biologists as well as environmen-

---

[3] Chargaff 1976.

talists, have felt that there is a perspective here worth considering. Perhaps it can contribute to a re-thinking of whether, when, and how inter-species genetic engineering should be used.

Arjun Makhijani
Takoma Park, Maryland
April 2001

Acknowledgements

I would like to thank Brent Blackwelder, President, Friends of the Earth (USA), for being open to these ideas and encouraging me to pursue them; Martha Herbert, Pediatric Neurologist at Harvard Medical School, who provided me with advice, many reference materials, and much intellectual stimulation; and Professor Richard Strohman, emeritus, University of California, Berkeley, who took the time to read this essay, comment on more than one draft, discuss it with me at length, and provide me with reference materials.

Izja Lederhendler, of the National Institute of Mental Health, provided me with important comments on several points, including behavioral adaptation. Marty Teitel, Executive Director of the Council for Responsible Genetics, provided a crucial philosophical criticism that led to a reformulation of a part of this essay. I am indebted to Professor Niall Shanks, of East Tennessee State University, who spent a great deal of time reviewing various drafts of this monograph, discussing it with me, and providing me with a very valuable education on many crucial points of evolutionary biology. Of course, as the author, I alone am responsible for the contents of this monograph, including any errors that it might contain.

Lois Chalmers, IEER's librarian, helped extensively with bibliographic research and proofing of the manuscript. I want also to thank all the supporters of IEER who send unrestricted contributions, which make projects such as this possible.

Finally, the sweetest acknowledgements, which are to my daughters. I want to thank Natasha, who repeatedly urged me to convert the notes and scattered ideas I had been talking about for some time into a finished piece. Her confidence in me and her encouragement gave me much of the motivation to do so. And I'm grateful that Shakuntala taught me to watch and appreciate the ants in our yard.

—Arjun Makhijani

CHAPTER 1: The Ecosystem in Us

As the jaguar searches the forest for prey at dusk, her spots camouflage her well. This commonplace of evolutionary adaptation hides a crucial question. How do the light and dark patterns of the forest transform themselves into the coat of the jaguar? The development of patterns of spots by retention of genetic mutations suited for survival may explain how evolutionary *change* occurs, but it provides no clue as to genetic structure itself.[4]

What is the correspondence between the internal biological structure and the external world and how is it expressed?

Consider another example from a myriad that could be put forth. The baby crocodile, recently emerged from its egg, lunges out of the water and accurately snaps up an insect. But it will not touch one that is immobile on the water surface or one that has already drowned and is sinking to the bottom.[5] How did the hatchling crocodile "know"—in the sense of instinctual

[4] Polanyi has made a similar observation about natural selection. "Natural selection tells us only why the unfit failed to survive and not why any living beings, either fit or unfit, ever came into existence." Polanyi 1958, p. 35. Levins and Lewontin have noted that "Early evolutionists did not take up the problem of the origin of life as a central issue. In *The Origin of Species* Darwin mentioned the problem only in passing and then metaphorically as the 'primordial form, into which life was first breathed.'" Levins and Lewontin 1985, p. 46. Charles Darwin's own goal was to explain changes and adaptation—that is the origin of *species*. It was, he wrote, "of the highest importance to gain a clear insight into the means of modification and coadaptation."—Darwin 1998 edition, p. 20. While this essay does not address the origin of life, it takes up the related questions of its nature and structure.

[5] Davenport et al. 1990.

grasping of the essence of external reality—from the moment of birth that live insects are food or that drowned dead ones sink and should be left alone? One may pose the question differently. What is the correspondence between the internal biological structure and the external world and how is it expressed?

Need-knowledge

I use the terms "know" and "knowledge" in a biological rather than anthropomorphic sense. A new noun and verb are needed to distinguish biological knowledge—which enables living beings to act to survive, adapt, and reproduce—from mechanism, which essentially denies biological knowledge, and from conscious knowledge of the intellectual variety. The terms "recognize" and "recognition," which can be more clearly attributed to living beings generally, cover the phenomenological aspects of the terms we need, but not those of the internal biological structure. Biological "knowing" of the external world corresponds to the internal urges that all living beings feel as their *needs*, which are lacks that must be satisfied for continued existence. I will call this "need-knowledge," which has both genetic and non-genetic aspects.

The actions that a living being performs for survival (such as eating, breathing, excretion, reproduction, flight from danger, and, in many species, nurture) necessarily relate that living being to its environment. There is, therefore, an internal biological structure that enables each living being to carry out just those activities (though with variations in efficacy). Picking up scents, seeing, leaping and grabbing the insect prey, fleeing from the predator, reproducing—none of these actions could possibly occur unless the internal biological systems of the various living beings were constructed so as to respond systematically to external events or to initiate them.

To return to the first example, the coat of the jaguar results from the internal reproductive system of the species. Indeed, coat pigmentation patterns arise directly from the genetic make-up of the jaguar. One may conclude, therefore, that one aspect of the jaguar's genetic structure is that it is a specific biological expression of the patterns of light and dark in the forest (integrated over time). In this sense the structure of the jaguar's genome contains particulars about *its* environment in forms that enable it and the species to survive.[6]

A part of the jaguar's genetic structure is one biological expression of the patterns of light and dark in the forest.

The second example shows a somewhat different aspect of genome-ecosystem relationship. In this case, the communication of the particulars of the insect to the crocodile is mediated by thc environment, which is external to both. The recognition by the crocodile of the insect as food comes via the sounds, sights, and smells which *represent* the insect externally to the crocodile. To grasp the insect as food, the

[6] I use the term "genome" to mean the full ensemble of material potentially involved in reproductive inheritance. Besides nuclear DNA, this may include DNA found in cytoplasm. The exact nature of this ensemble of material does not affect the broad hypotheses discussed here, though of course, it would affect their further exploration.

crocodile need only integrate the insect's various simultaneous appearances and convert them into an internal signal that food is present. The genetic structure of the crocodile only needs a biological expression of the phenomena, rather than of the insect, as such, in its entirety.[7]

The common element in both examples is that the external environment needed by a living being has an internal genetic expression. The basic hypothesis about the genetic-environment correspondence is that the *genome of any species is the internal biological expression of the ecosystem needed by that species for its*

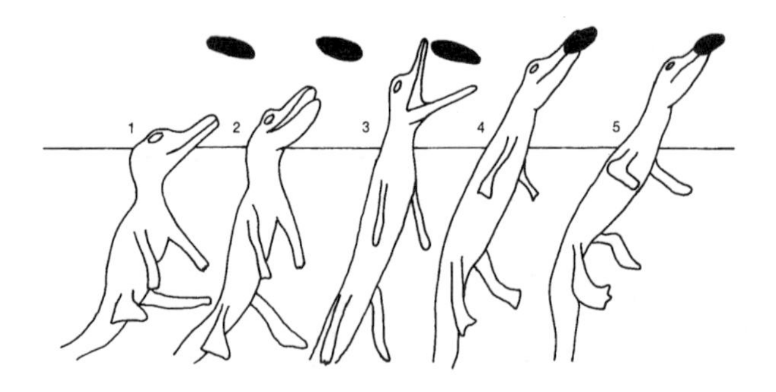

A hatchling crocodile has enough need-knowledge of its ecosystem to enable detection of live insects (represented by black ellipsoids) as food. Diagram source: Davenport et al. 1990. p. 575. Reprinted with the permission of Cambridge University Press.

[7] The sexual partner of the insect needs, and has, a more complete internal representation and a more detailed need-knowledge of the phenomena.

existence.[8] In other words, the genome of a species has an internal structure that corresponds to the specific ecosystem that it needs to survive both individually and intergenerationally.[9] This is not to say that the genome is the exclusive internal expression of the ecosystem. The environment in which the fertilized cell develops and the external environment after the individual is born are also crucial to shaping living beings.[10] The first aspect involves the environment prevalent during the process of reproduction. The second is the post-natal aspect, which involves the environment in which individuals conduct the day-to-day business of living. There are large variations in the survival of trees during storms that depend not only on genetic variations between species

The genome of a species has an internal structure that corresponds to the specific ecosystem that it needs to survive.

[8] The internal genomic expression of the ecosystem that a species needs being is not a "snapshot" of the external ecosystem at any particular time. Moreover, since there is a wide scope for non-genetic adaptation and similar environments in many places, the ecosystem needed by a species as expressed in its genetic structure may be geographically more variable than the specific context of the evolution of that species. See below for further discussion.

[9] Richard Lewontin has made the following general observation about the relationship between environment and organism: "The environment of an organism is the penumbra of external conditions *that are relevant to it* because it has effective interactions with those aspects of the outer world." Lewontin 2000, pp. 48-49, (emphasis added).

[10] Of course, the cellular context of the development of the zygote into an individual is also central. For instance, in the process of cloning the famous sheep Dolly, it was necessary to place the parent nucleus (derived from an udder cell) into an enucleated oöcyte, and not any other cell. Cloning shows the essential role of the cytoplasm of the öocyte in the process of cell differentiation needed for a zygote to become a fully developed individual, at least in mammals.

and within tree species, but also on a host of environmental and developmental factors that characterize the specific history of each tree. But over the long-term, the essentials of day-to-day living must be expressed in genetic structure. A tree species cannot survive if the range of windspeeds that it would typically experience were not incorporated, through the motions that wind has induced in trees over the ages, into that aspect of its genetic structure that gives its trunk and branches the tensile and shear strength to survive most storms.

A tree species cannot survive if the range of typical storm windspeeds were not incorporated into the part of its genetic structure that makes its trunk and branches strong.

CHAPTER 2: Incorporation and Excorporation

The actions that relate living beings to their environment can be grouped into two very broad categories that throw light on the genome-environment relationship. The first is *incorporation*—the process by which an individual internalizes parts of the environment in some way as part of the process of living. To paraphrase the famous physicist Richard Feynmann, yesterday's lentils are today's muscle tissues or sperm cells. The oxygen in the air when breathed in becomes part of hemoglobin in blood; carbon dioxide in the air turns to carbohydrates in grasses.

Yesterday's lentils are today's muscles or sperm cells.

Incorporation is also part of the process of reproduction. For instance, reproduction by mitosis requires incorporation of nutrients. In sexual reproduction, sperm is incorporated into the oöcyte to form the cell that becomes the new individual. More broadly, incorporation also includes internalization of social structure needed for survival.

The obverse set of processes can be grouped under the term *excorporation*. Carbohydrates once used up inside the body become carbon dioxide and other unneeded molecules to be excreted. Excretion of bodily chemicals involves a variety of functions such as getting rid of toxins or attracting a mate. Giving birth is an act of excorporation. And "behavior" in all species having social structure can be understood as excorporation of internalized social structure.

Excretion of bodily chemicals gets rid of toxins or helps attract a mate. Giving birth is another example of excorporation.

The internal structure of the insect appears to the crocodile through excorporation of light, chemicals, and sounds by the insect. Similarly, the flight of the insect away from its predator is mediated by the ecosystem through excorporation by the crocodile. There must therefore be a systematic correspondence between the internal genomic structure of the prey and predator that allows their relationships to exist. In other words, in so far as the predator and prey are related, their internal genomic structures are two expressions of the same ecosystem structure. Of course, one is the prey and the other the predator, and there are also essential differences in their internal structures. In sum, incorporation and excorporation are approximately reciprocal sets of activities that are needed to perpetuate a species as such and the individuals within it. Genetic structure is, of course, but one aspect of the matter.

In so far as the predator and prey are related, their internal genomic structures are two expressions of the same ecosystem structure.

The acts of incorporation and excorporation, in their totality, constitute the fundamentals of living. They can even be used to define life. *A living being, by its internal structure, needs to incorporate its environment and to excorporate into its environment so that it may go on existing as such.*[11] To put it more

[11] This is similar to, but also somewhat different from, the autopoiesis description of Fritjof Capra (Capra 1996, Chapter 7). The definition of living being given here allows us to create a functional boundary between the living and non-living, though, as Lewis Thomas observed "we [human beings] are shared, rented, occupied" for instance by mitochondria, " and in a strict sense they are not ours." He concluded that "…I am grateful for differentiation and speciation, but I cannot feel as separate an entity as I did a few years ago…." Thomas 1974, pp. 4-5. J. Scott Turner has

simply, living beings have *needs* that relate them to their environments; inanimate things do not. This is the central feature that distinguishes the living from the inanimate. One might modify the famous Cartesian claim "I think, therefore I am" and say instead: I eat therefore I am—and vice versa.[12]

Such a construct is supported by genetic research. Genes that code for proteins directly associated with the vital functions of incorporation and excorporation tend to be conserved across species. But it is difficult to say what detailed inferences about entire genomes or specific genome sequences whose functions are not yet understood may be drawn from observations of their patterns of incorporation and excorporation, given that present ideas about genetic structure are the result mainly of the dissection of DNA.

One might modify the famous Cartesian claim "I think, therefore I am" and say instead: I eat therefore I am—and vice versa.

---

published an extensive analysis of this fuzziness of boundaries between living beings and their environment, providing examples such as the way a cricket uses a leaf to improve the efficiency of the sound it needs to produce for its mating call and the creation of burrows by earthworms as they ingest the soil to extract organic matter from it. Turner 2000, p. 165 and pp. 115-119.

[12] This is evident if one focuses on what French people do, instead of on what their philosophers say. Once we have identified that which is common between humans and other living beings—they all have needs they must fulfill in order to go on existing as living beings—we can begin to identify what makes humans different. For instance, in growing from babies, human beings acquire the capacity to go on a hunger strike—that is, negate a biological need—in order to secure what is precious. How such a notion of freedom can be elaborated so as to throw light on *human* nature is beyond the scope of this essay. (These ideas were printed in a letter to the editor on May 17, 1997, sent to the *New York Times* by the author as a comment on the essential differences between Deep Blue, a chess playing computer, and human beings.)

CHAPTER 3: Modes of Expression

If the genome of any species is an expression of the ecosystem it needs, then there must be specific ways in which aspects of the ecosystem that that species has experienced during its evolution gets expressed within genetic *structure*, which consists both of the substance of the chemicals and their form (their specific shapes). This marriage of substance and form in structure expresses internally the aspects of the external world needed by the organism. For instance, hemoglobin is, in part, the internal expression of oxygen—in fact oxygen must physically fit into the structure of the hemoglobin protein molecule. The fit is so good that, at typical lung air pressure, hemoglobin is 98 percent saturated with oxygen.[13] For a large variety of animals, the part of the genetic structure that produces hemoglobin can therefore be viewed as one expression of the oxygen component of the ecosystem. That same genetic structure also expresses carbon dioxide, which hemoglobin transports out of the body via the lungs, as well as nitric oxide, which regulates blood pressure.

Hemoglobin is, in part, the internal expression of oxygen, since oxygen molecules must fit into the structure of the hemoglobin protein molecule.

We might also consider the correspondence between the genetic structure of chloroplasts and mitochondria, which are internal to cells, and the carbon dioxide and oxygen in the atmosphere.

[13] Campbell 1996, p. 846.

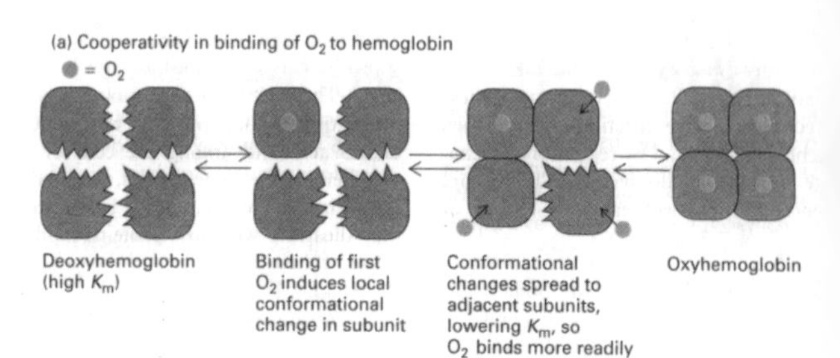

Hemoglobin: Incorporating oxygen. FROM: MOLECULAR CELL BIOLOGY by H. Lodish, A. Berk, S.L. Zipursky, P. Matsudaira, D. Baltimore, and J. Darnell © 1986, 1990, 1995, 2000 by W. H. Freeman and Company. Used with permission.

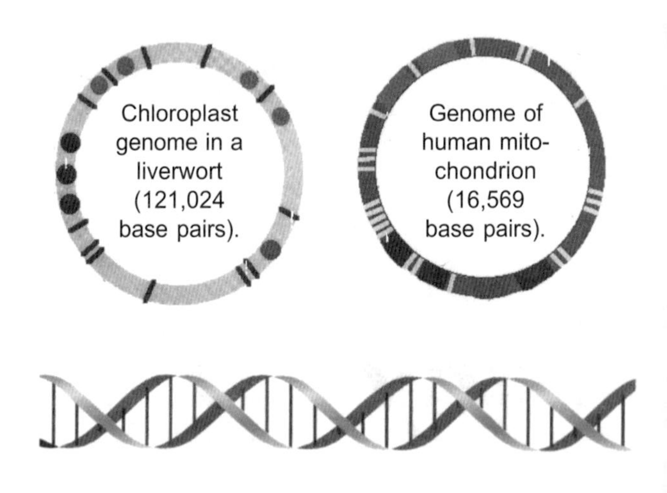

The circle-shaped DNA in mitochondria and chloroplasts is essential to the carbon-oxygen cycle and hence to reproduction of the double-helix-shaped nuclear DNA. Sources for circle-shaped DNA: John W. Kimball; for double-helix: FROM: MOLECULAR CELL BIOLOGY by H. Lodish, A. Berk, S.L. Zipursky, P. Matsudaira, D. Baltimore, and J. Darnell © 1986, 1990, 1995, 2000 by W. H. Freeman and Company. Used with permission.

Chloroplasts, which have their own DNA, are the reducing agents in plant cells that convert sunshine, carbon dioxide, and water to carbohydrates and oxygen. Mitochondria, with a DNA structure similar to chloroplasts,[14] perform the reverse function by oxidizing carbohydrates to yield energy, water, and carbon dioxide. Were there no genetic structure within chloroplasts and mitochondria that allowed for recognition of carbon dioxide and oxygen, the overall oxygen and carbon cycles in Nature that are needed to sustain life simply could not exist.[15]

At another level, we might consider internal cellular differentiation, which occurs mainly through the differential expression of genes that are present in every cell. The sequence, rate, and entire environmental context, from the intra-cellular to the ecosystem level, in which genes are turned on and off determine the specific outcome of the cellular developmental process. In view of this, geneticist D. J. Pritchard has noted that gene expression can be viewed as an incorporation of the environment into the living being:

> "Gene expression is controlled directly or indirectly by the intra- and extra-cellular environments to which the DNA in which the genes are encoded is exposed, while components of the environment become incorporated into bodily structure as a feature of expression of the genes. Phenotype at all

---

[14] Both have a closed circle DNA structure. For illustrations, Kimball 2000.

[15] Childs 1996. For a philosophical discussion of mitochondria and humans, see Thomas 1974.

There is no simple one to one relationship between function, ecosystem feature, and genetic structure.

Jessie Cohen, Smithsonian's National Zoo.

The genes that shape the jaguar's coat are, in part, its internal expression of the patterns of light and dark in the forest.

levels is thus the product of interaction between the genome and the environment and either can be limiting."[16]

That there is no simple one-to-one relationship between function, ecosystem feature and genetic structure is obvious from the simple example of a single protein, hemoglobin, that fulfills many functions that interact with the functions of other genes (in maintaining and changing blood pressure for instance).[17] As

---

[16] Pritchard 1990, p. 151.

[17] See Strohman 1997 for a discussion of the role of many genes in single functions. The one-gene, one-protein model is, in any case, being discarded, partly due to indications that there may be far fewer genes in the human genome than previously thought and that this number is only about a factor of two greater than the number of genes in a roundworm's genome. See Gould 2001.

another example, the color of the jaguar's coat not only plays a role in camouflage, but also in the rates at which its body absorbs and radiates heat. Other aspects of its coat, texture and thickness, also play a role in its heat balance.

There are also many potential ways in which genes express a specific aspect of the environment, so that there is no necessary "best" answer to the problem of survival for living beings in a specific ecosystem. There is also considerable evidence of redundancy of function in genetic structure, indicating that the ecosystem of the organism is expressed in many ways within that organism even for single functions. Finally, if there are multiple potential genetic expressions of ecological realities, a profusion of species and subspecies becomes more understandable.

Evidently, the internal expression of the ecosystem in us is not a point-to-point map. Far from it. The same physical conditions have produced a large variety of organisms corresponding to them, as biologists Richard Levins and Richard Lewontin have observed.[18] One can understand this from an informational standpoint. For instance, much of the prey's specific internal structure is irrelevant to the predator. This allows a variety of insects to be expressed in the same internal biology of the predator, since, for the predator, the essence of the prey is its catchability

---

[18] Levins and Lewontin note that "[t]he consequence of the interaction of gene, environment, and developmental noise is a many-to-many relationship between gene and organism. The same genotype gives rise to many different organisms, and the same organism can correspond to many different genotypes." Levins and Lewontin 1985, p. 94. See also Lewontin 2000, pp. 53-68.

and its digestibility. The crocodile needs to sense that the insect is edible and flies. This raises the possibility that there may be many different genotypes that could make use of essentially the same external phenomena to grasp the insect as food. And, indeed, there are—birds and crocodiles, for example.

Many different genotypes make use of essentially the same external phenomena to grasp the insect as food—birds and crocodiles, for example.

Further, in complex living beings, organs are the internal expression of a living being's ecosystem beyond the level of single cells or gene-protein relationships. Both excorporation and incorporation occur via the organs, which are the internal biological instruments of the genome-ecosystem where the instinctual grasping actually takes place. More than that, organs are also among the principal locations of non-genetic adaptation to the environment that necessarily occurs in the process of living.[19]

Non-genetic adaptation, which is one aspect of what one might call need-learning, results in non-genetic need-knowledge, and is as fundamental to survival as genetic adaptation.[20] The incomplete representation of the essence of the external by integration of a set of phenomena not only produces vulnerabilities, for instance,

---

[19] As a speculative corollary of this line of thought, one might ask whether the folded structures that characterize genetic and protein structure may be the result of convolution processes by which the external environment becomes expressed in internal living structures.

[20] Edward Goldsmith has called living beings "intelligent" in this sense of adaptation. See Goldsmith 1998, Chapter 32. However, as in the case of biological knowledge, biological adaptive learning should be distinguished from the processes by which human society creates printing presses and books. There is ample evidence that adaptive learning and the consequent adaptive, non-genetic, need-knowledge exists in

to parasites, but also enables a variety of external phenomena within broad parameters to be represented in the same way. Many different insects can be represented internally in the same way as food, without much special regard to the differences between various species. This property allows specific species to adapt to new environments (within limits) without genetic change. The individual crocodile in a new location will learn to recognize the different species of insects that are present. This plasticity of response does not arise from genetic change, but from the fact that the relationship of an individual's genetic structure to the external is mediated by internal systems (organs) that are themselves adaptive.

. . . .

Non-genetic adaptation is as fundamental to survival as genetic adaptation.

Complex beings also adapt by changing behavior. The brain seems to have evolved in them as an organ that expresses both the external ecosystem structure as well as internal biology. It mediates and regulates relations between the two. In this view, the brain is a specific internal biological expression of external survival needs, which necessarily have social, including behavioral, aspects. While change in behavior involves internal changes, in

---

non-human species. Consider, for instance, the capuchine monkeys of Venezuela who, during peak mosquito season, know how to find and apply mosquito repellent. They collect a species of millipede rich in benzoquinones, which are powerful mosquito repellents. They crush the chemicals out of the millipedes and apply them to their bodies. The monkeys don't have knowledge of benzoquinones in the Cartesian intellectual sense that the researcher who studied them does, but they nonetheless grasp the essentials instinctually, and with fine enough timing to be able to ward off mosquitoes effectively at the peak of the mosquito season. Angier 2000, p. D5.

Order, Information, and Need

Physicist Erwin Schrödinger, in his essay *What is Life?*, noted that the process of living was the reproduction of "order based on order"—ordered genetic structures reproducing themselves by the process of living.[21] The creation of order is a reduction of entropy and is understood as such in the second law of thermodynamics in both its thermodynamic and informational interpretations. The maintenance of low entropy—that is a high degree of order—using energy mainly from the sun is a central physical characteristic of the reproduction of ecosystems by living beings. Some of the defining aspects of the low-entropy state that characterize living beings are in genome-ecosystem *relationships*, which depend on internal structures, like brains and bladders, needed to establish them, as well as physical phenomena like sound waves and chemicals like pheromones, that mediate them.[22] The crocodile needs far less information about the insect than the mate of the insect—and mates can gather more information about each other because their genetic structures are very close. In this way, the expressions of genetic structure serve, in part, as differential filters for information according to need. Is the need, for instance, for a mate or a meal? Of course, there are also non-genetic aspects of information filtering.

---

[21] Schrödinger, 1967 edition, p. 68 and Chapter 5, more generally.

[22] For a discussion of orderliness, disorderliness, and entropy that includes the environment of the organism, see Turner 2000, pp. 11-25 and pp. 116-119.

what is remembered, for instance, it may or may not involve development of organs. In other words, behavior provides an additional level of plasticity—that is, the capability for behavioral adaptation to new phenomena without concomitant genetic change. Developmental and behavioral factors mean that genetic structure is, within limits, shielded from many kinds of environmental change. However, at any time different species are shielded in different ways and to different extents, with the result that adaptation by species at all levels is occurring simultaneously within an evolving ecosystem as a matter of course.

The brain seems to have evolved as an organ that expresses external ecosystem structure as well as internal biology.

We may deduce from the above that the genome of a species does not represent a "snapshot" of the external environment as it exists at any time. First of all, genetic change is not instantaneous. Secondly, as we have noted above, it does not need to be, since there is substantial non-genetic adaptability in organisms. Our muscles and livers and brains can and do change in response to developmental and environmental factors without corresponding genetic change. As a corollary we may postulate that the collection of species that exists at any time is not genetically contemporaneous with the external environment. This has some implications for the reproduction of ecosystems and for evolution.

CHAPTER 4: Reproduction of the Ecosystem

Living beings do far more than reproduce their own species. The reproduction of the immense and complex relationships in nature could not possibly occur through ecosystem relationships that were incidental to the reproduction of species. Flowers and bees could not be involved in a *system* of mutual survival and reproductive success unless each contains in its genetic structure some expression of the other. As we have discussed, this expression is not direct, but is mediated by sets of appearances of both in the environment external to them that represent the essence of the one living being to the other.[23] It is this environment-mediated internal structure in both flowers and bees that has evolved over time, along with the environment.

Ecosystems as such tend to be actively reproduced by the living beings within them.

Ecosystems contain the ensemble of temporarily living and temporarily non-living matter in constant systemic exchange in all four permutations of those two categories.[24] This complex of system of exchange can only be sustained if ecosystems *as such* tend to be actively reproduced by the living beings within

[23] The terms "external" and "internal" are defined by the acts of excorporation and incorporation of living beings. Evidently, there are multiple and overlapping layers of internal and external. See discussion in footnote 11 above.

[24] There are transformations of the living into the living, the living into the non-living, the non-living into the living, and the non-living into the non-living.

them. That is, the collective acts of a species must not only enable the immediate survival of a species they must also systematically contribute *to the reproduction of the ecosystem that that species needs.* This is a crucial aspect of species survival. Material cycles (oxygen, nitrogen, and carbon, for instance) in nature cannot be (more or less) closed unless the genomes of the living beings within the ecosystem collectively contain sufficient need-knowledge of the ecosystem to contribute to its reproduction. However, the system is never exactly closed, given that changes in entropy and time are unidirectional and that both the ecosystem and the species in it are evolving (see below). For the same reason, reproduction of an ecosystem is never exact, but is rather a tendency of the system to maintain certain kinds of flows through it and structures within it.

If particular species contribute to the reproduction of their specific ecosystems and all species together tend to reproduce the global ecosystem,[25] the hypothesis that species evolve to occupy pre-existing "niches" becomes very questionable. Richard Lewontin has observed that:

> "The concept of an empty ecological niche cannot be made concrete. There is a non-countable infinity of ways in which the physical world can be put together to describe an ecological niche, nearly all of which would

[25] Goldsmith calls the global ecosystem the "ecosphere." Goldsmith 1998, Chapter 19.

seem absurd or arbitrary because we have never seen an organism occupying such a niche." [26]

Rather, the ecological niche should be understood as "a space defined by the activities of the organism itself."[27] *Each species systematically contributes to the survival of the ecosystem it needs because that is the very process by which specific species actually evolve and come into being.* To put Lewontin's observation in the context of the genetic hypotheses in the present essay, the ecological "niche" and the genome of the species come into existence as part of the same process—that is, they co-evolve.

The collective acts of a species must systematically contribute to the reproduction of the ecosystem that that species needs.

Even though the reproduction of ecosystems and genomes occurs mutually, no single species contains the genetic-need-knowledge to internally represent the entire ecosystem. Even a species' own ecosystem is incompletely represented internally. The insect that is the crocodile's food is internally represented in the crocodile only via the integration of a limited set of appearances that are *normally*, but not always, adequate.

A heron scooping up a fish in its beak may not detect the parasite in its prey, or at least not detect it well enough to prevent becoming, on occasion, a prey itself.[28] One central consequence of this partial and incomplete representation is

---

[26] Lewontin 2000, p. 49.

[27] Lewontin 2000, p. 53.

[28] For the complex interactions between fish, parasites, and herons, see, for instance, Spalding and Forrester 1991.

Pam Gardner; Frank Lane Picture Agency/CORBIS

Tension: The heron's uncertainty about what's in the fish enables parasites to flourish.

Global-scale symbiosis is essential to all species, since no species can reproduce the ecosystem it needs. Competition should be seen within that symbiotic context.

that *no single species can by itself reproduce the ecosystem it needs.*

The hypothesis of the collective reproduction of the global ecosystem by the species within it is conducive to the idea of evolution through symbiosis. In fact, the entire process of living through incorporation, recognition, and excorporation that results in the reproduction of species and ecosystems can be viewed as a large-scale symbiosis. Global-scale symbiosis is essential to all species, since no species can reproduce the ecosystem it needs.[29] Competi-

[29] The hypotheses discussed in this paper can be more explicitly linked to the *Gaia* hypothesis by defining a time-dependent global ecosystem "genome" that corresponds to the reproduction of the evolving ensemble of the living beings on Earth. A historical description of the concepts involved in the *Gaia* hypothesis can be found at http://www.magna.com.au/~prfbrown/gaia_jim.html. See also Goldsmith 1998, and Capra 1996. A number of scientific papers on the subject of symbiosis and evolution can be found in Margulis and Fester, eds. 1991.

Appearance, Essence, and Uncertainty

Jean-Paul Sartre, in *Being and Nothingness*, wrote that the essence of a thing is not hidden in its interior as a secret to be revealed if its outer layers are peeled off. Rather, the essence of something "is the manifest law that presides over the succession of its appearances, it is the principle of the series."[30] However, Sartre did not relate how the essence of something is conditioned by the other party to the series of appearances—the living being to whom the thing appears, who has a vantage point and an internal structure of needs. The essence of something that is grasped by another is the principle of a series of appearances that unifies subject and object in a specific, defining relationship. In fact, appearances need not define a single essence, since the manifold appearances of something can potentially be grouped according to many principles. As we have discussed, whether an insect is a mate or a meal depends both on the insect and the (engaged) observer of the insect. For the female black widow spider, its male counterpart is both, in sequence. Moreover, both parties to the appearance are changing—indeed, they are both changed by the process. Finally, the number (and/or duration) of appearances is necessarily finite. As a result of these factors, we may conclude that uncertainty is inherent in relationships within ecosystems, since they depend on the communication of essence through appearances.[31] This has important implications for our ability to predict the problems that might arise from genetic engineering (see Chapter 5).

[30] Sartre 1966 edition, pp. 5 and 6.

[31] These constructs are evidently related to Heisenberg's uncertainty principle in physics. See Tautz 2000 for discussion of uncertainty in relation to genetics.

tion should be seen within that symbiotic context.

Since there are complex disjunctions between the global ecosystem and the species that exist at any time, symbiosis is never perfect and never static. One might view this disjunction as one of the driving forces in evolution.[32] For instance, the fact that parasites residing in prey are insufficiently represented in the internal biology of predators to be detected enables parasites to flourish. The tension between parasite and host also raises the possibility, in specific instances, of the evolutionary transformation of parasitism into symbiosis. For instance, mitochondria, which are essential to cell metabolism, may have begun as bacteria that invaded larger single-celled organisms long ago.[33]

All species together tend to reproduce the global ecosystem.

Finally, living beings, collectively, do not reproduce the ecosystems they need without

---

[32] When responding to objections to his theory, Darwin noted this disjunction: ". . . organic beings . . . are not as perfect as they might have been in relation to their conditions....Nor can organic beings, even if they were at any time perfectly adapted to their conditions of life, have remained so, when their conditions changed, unless they themselves likewise changed . . ." Darwin 1998 edition, pp. 263-264. Richard Levins and Richard Lewontin have noted that, despite Darwin's remarks about "perfection of structure and coadaptation," his view was that adaptation produces living beings that "tend to progress toward perfection." Levins and Lewontin 1985, p. 26-27.

[33] This thesis was put forward in 1918 by a French biologist Paul Poitier—see Hubbard and Wald 1993, p. 164. This book also provides a discussion of mitochondrial DNA in human beings, pp. 163-167. For a more general discussion see Wesson 1991, pp. 157-167.

Buddy Mays/CORBIS

Black widow spider: It grasps the essence of its male counterpart as a mate and a meal, in sequence.

fail. Besides the disjunction between the internal structures of living beings and the global and local ecosystems they inhabit, the biogeosphere is subject to forces beyond the control of the living beings in it. Further, the rates of some of the changes that occur in the biogeosphere far exceed the possible rates of adaptation by specific species that might exist at any specific time. Great and/or sudden changes in ecosystems due to large volcanic eruptions or asteroid impacts are examples of such natural forces. But despite the enormous and sudden changes in the Earth's environment in the past, life, in its essential characteristics, has persisted not only because survivors could adapt to environmental changes, but also because they could shape the new environment to life's needs. Even volcanic ash soon becomes part of the stuff of life. But that does not ensure the survival of specific species in the face of drastic and rapid environmental change. After all, most species that once existed are now extinct.

CHAPTER 5: Genetic Engineering and the Environment

Human beings are now busy producing some of the most rapid changes ever known in the global ecosystem. The widespread use and misuse of antibiotics has already caused the emergence of resistant bacterial strains. Emissions of large amounts of greenhouse gases are causing as yet ill-understood changes in the in the patterns of circulation of energy and materials through various ecological cycles. Research on hormonally active pollutants (such as dioxin) suggests that they can change the expression of genes during fetal development and afterwards as well.[34] The questions of what overall genetic and ecosystem changes these types of human activities may induce and at what speed those changes might occur have hardly begun to be systematically posed.

The question of what overall genetic and ecosystem changes human activities may induce has hardly begun to be systematically posed.

Still, the widespread misuse of antibiotics and the introduction of many mutagenic and carcinogenic chemicals have provided ample proof that human activities can produce unintended, rapid, and possibly disastrous genetic change. In other words, severe genetic and environmental changes have already occurred due to the introduction into the environment of molecules that are far simpler than genetically engineered plants.

---

[34] Colborn, Dumanoski, and Myers, 1996, pp. 40, 120, and 203-207.

If the hypotheses discussed in this essay are even plausible, then we must conclude that the introduction of new forms of life into nature by inter-species genetic engineering may, at least in some cases, induce changes in ecosystems that may be substantial, unpredictable, and possibly even drastic. Inter-species genetic engineering, which uses genetic materials from widely different species, suddenly creates new genomic structures, each of which will try to create an external ecosystem structure that corresponds to an internal need-structure that is to some extent unknown. Radically new forms of life that mix genes from widely differently groups of living beings (fish genes in plants for example) could create new pressures on ecosystems.[35]

Inter-species genetic engineering suddenly creates new genomic structures that will try to create an external ecosystem structure that is to some extent unknown.

A hint comes from research on Bt corn and monarch butterflies. (Bt stands for *bacillus thuringiensis*, the bacterium that carries the gene that is engineered into the corn.) Bt corn contains a gene from *bacillus thuringiensis* so as to enable it to internally produce a pesticide, thus obviating the need for external pesticide application. In an experiment, Cornell University scientists applied pollen from Bt corn to milkweed, some of which grows near cornfields and placed monarch butterfly caterpillars on them, since they feed on milkweed leaves. After four days, 44 percent of the monarch larvae that fed on Bt corn leaves died, while none of the

[35] Plants would not necessarily "try," in the biological sense, to create some aspect of a fish ecosystem, since the expression and function of genes is generally contextual—that is jointly determined by combinations of genetic and non-genetic structure specific to particular living beings and to their environment.

Defining genetic engineering

It is necessary, when considering restrictions on genetic engineering, to define what we mean by "widely different" genomic structures, since some hybridization, for instance with plants, has long been carried on by non-genetic engineering techniques. A place to start might simply be to give primacy to natural reproductive recognition. Genomic structures that are reproductively compatible through macroscopic techniques, such as grafting, allow natural biological recognition to determine the genetic modifications that can be carried out. Amory and Hunter Lovins have described the distinction thus: "Traditional agronomy transfers genes between plants whose kinship lets them interbreed. The new botany mechanically transfers genes between organisms that can never mate naturally: an antifreeze gene from a fish (Arctic flounder) rides a virus host to become part of a potato or a strawberry."[36]

caterpillars on the control leaves that had no pollen or that had corn pollen from non-engineered plants died.[37] While the ecological significance of this experiment is not yet clear, it is evident that genetically engineered corn has been introduced on a vast scale without sufficient consideration of its effects on ecosystems. If it can adversely affect monarch butterfly caterpillars so severely, how many other types of

[36] Lovins and Lovins 1999.

[37] Losey, Rayor, and Carter 1999.

Texas Department of Agriculture

Bt or not Bt? A monarch butterfly's caterpillar may not be able to avoid the danger.

flora and fauna might it also affect? We scarcely know.

If history is any guide, some of the nastiest changes will come as surprises—as, for instance, that some artificial chlorinated chemicals behave like estrogens—because we do not adequately understand the complex correspondence of ecological, chemical, and genetic structures. As a result, we cannot realistically assess all essential aspects of the safety of genetically engineered foods for human beings, or even decide which aspects are essential for the long-term.[38]

There is even less understanding about what genetically engineered plants and animals may do to the environment. Furthermore, many

[38] For a discussion of the lack of knowledge of the toxicological effects of genetically engineered food, see, for instance, Millstone, Brunner, and Mayer 1999.

changes may be undiscoverable by practical means until it is much too late to stop the damage. Diethard Tautz has suggested an "uncertainty principle" in genetics: for "genes or genetic functions that have only a very small effect on the fitness of an individual, but are nonetheless important for long-term fitness within a population," an adequate understanding may require "experiments that involve the whole population of the respective species."[39] Of course, this means that nearly the entire population would have to be changed to discover whether deleterious changes have occurred—a genetically self-defeating proposition.

The terrible surprises that genetic engineering may hold in store were dramatically demonstrated by an Australian effort to genetically engineer the mousepox virus.

The terrible surprises that genetic engineering may hold in store were dramatically demonstrated by an Australian effort to genetically engineer the mousepox virus (which is related to the virus that causes smallpox in human beings but does not attack humans) in order to control rodent population and reduce crop damage. The mousepox virus was modified by the insertion of a gene associated with the control of the production of interleukin-4, an immune-system-related molecule. The goal of the genetic engineering was to increase the immune response of the rodents so greatly that the eggs of the mice would be rejected as foreign objects, in the way that external disease-causing agents are attacked by the immune system.

The outcome was, in more than one way, the opposite of what was expected. Instead of

---

[39] Tautz 2000.

strengthening the immune system, the genetically engineered virus suppressed it. Instead of being less lethal, the new virus was more lethal. Mortality was high even among vaccinated mice and mice bred for resistance to mousepox. In other words, the engineered virus not only defeated primary immune response, it also inhibited "the expression of immune memory responses."[40] Surviving mice were permanently disabled.

The mousepox experiment's outcome was the opposite of what was expected. Instead of strengthening the immune system, the genetically engineered virus suppressed it. Surviving mice were permanently disabled.

These effects were observed during 1998 and 1999. But public discussion of them was suppressed for some time, for fear of spreading the information that it is relatively easy to create a new lethal virus—one that can even defeat vaccinations. Debate has so far centered on the potential of the technique for the deliberate creation of deadly new biological warfare agents for which no timely responses may be possible. How does one balance such possible risks of genetic engineering experiments, which unintentionally illuminate a path to deadly diseases and biological warfare, with any supposed benefits? And what about accidents arising from research, to say nothing of genetically engineered organisms now being introduced into the environment far beyond the laboratory?

The questions arising from the mousepox research are made more troubling by the fact that little independent research has been carried out on crucial ecological aspects of genetic engineering. In a review of the existing literature on the ecological effects of genetically

---

[40] Jackson et al. 2001. See also Broad 2001, p. A8.

engineered organisms published in *Science*, L.L. Wolfenbarger and P.R. Phifer concluded that "key experiments on both environmental risks and benefits are lacking."[41] Specifically, the "ecological consequences in nonagricultural habitats and ecosystems" of genetically engineered organisms "remain largely unstudied" despite indications of risk from past experience with crops that are crucial to the world's food supply:

> "No published studies have examined whether introgression of transgenes or its potential ecological consequences have occurred in natural populations; however, past experience with crop plants suggests that negative effects are possible. For seven species (wheat, rice, soybean, sorghum, millet, beans, and sunflower seeds) of the world's top 13 crops, hybridization with wild relatives has contributed to the evolution of some weed species. In some cases, high levels of introgression from cultivated or introduced relatives have eliminated genetic diversity and the genetic uniqueness of native species, effectively contributing to their extinction."[42]

Like Tautz, Wolfenbarger and Phifer note that some effects cannot be determined from small-scale experiments:

> "Unknown risks may surface as the frequency and scale of the introduction increases. Because some consequences, such as the

The "ecological consequences in non-agricultural habitats and ecosystems" of genetically engineered organisms "remain largely unstudied."

[41] Wolfenbarger and Phifer 2000, p. 2088.

[42] Wolfenbarger and Phifer 2000, p. 2088.

> probability of gene flow, are a function of the spatial scale of the introduction, limited field experiments do not always sufficiently mimic future reality prior to widespread planting."[43]

The rub is that if it is necessary to resort to widespread planting to discover adverse effects, then it will probably be too late to do anything about the harmful effects that are discovered. As long ago as 1976, when biotechnology as a large economic prize was only a gleam in the eyes of researchers, biologist Erwin Chargaff, in a letter to the journal *Science*, pointed out that "you cannot recall a new form of life."[44]

The unwanted dispersion of StarLink corn illustrates biologist Erwin Chargaff's warning: "you cannot recall a new form of life."

The example of the StarLink variety of corn should be taken as an early warning of this problem. StarLink is the trade name for a type of Bt corn containing the Cry9C protein. This corn variety had been approved for animal feed, but not for human consumption due to data indicating a potential for producing allergic reactions. Testing initiated by Friends of the Earth in the year 2000 showed that it was present in tortillas purchased in a supermarket.[45] As testing became more widespread, the estimates of the amounts of contaminated corn increased from 70 million bushels to 430 million bushels.[46] The latter figure represents enough food calories to supply many millions of people with grain for an entire year.

---

[43] Wolfenbarger and Phifer 2000, p. 2090.

[44] Chargaff 1976.

[45] FoE 2000

[46] Kaufman 2001a.

Even though StarLink was planted on less than 1 in 5,000 acres of land planted to corn, it is now difficult to be confident that corn is completely uncontaminated. A special definition of uncontaminated corn has had to be created: it is corn containing less than 1 kernel in 2,400 of StarLink.[47]

In early 2001, a portion of the U.S. seed supply of corn meant for human consumption was found to be contaminated. The U.S. government is buying back several hundred thousand bags of seeds.[48] In a practical recognition of the reality that StarLink corn cannot be recalled, the company that made it, Aventis, asked the Environmental Protection Agency to retroactively rule it safe for human consumption. It had not been certified fit for humans due to an unknown risk of allergic reactions.

Chargaff also noted that "[b]acteria and viruses have always formed a most effective biological underground. The guerilla warfare through which they act on higher forms of life is only imperfectly understood. By adding to this arsenal freakish forms of life—prokaryotes propagating eukaryotic genes—we shall be throwing a veil of uncertainties over the life of coming generations. Have we the right to counteract, irreversibly, the evolutionary wisdom of millions of years, in order to satisfy the ambition and the curiosity of a few scientists?"[49] Profits,

---

[47] Kaufman 2001a.

[48] Kauffman 2001.

[49] Chargaff 1976. Goldsmith has provided other examples of scientists raising similar questions. See Goldsmith 1998, Chapter 57.

of course, have since come to play a larger role.[50]

Chargaff alludes to the very slow rate of change of natural genetic-ecosystem interactions compared to the DNA shot-gun and other engineering methods that can suddenly create new genomic structures impossible in nature. This rapid change may present challenges and uncertainties as big as those from the fact of creating new structures. While species that evolved in one local ecosystem can often adapt to new places, the ecosystems into which new species are introduced do not necessarily have a reciprocal adaptability.

DNA shot-gun engineering can suddenly create new genomic structures impossible in nature.

The introduction of European rabbits into Australia in 1859 by a wealthy homesick landowner is a notorious example, estimated to be the "most destructive and most expensive transfer ever of an animal from one country to another." They have wreaked havoc on all forms of vegetation and overcome efforts to eradicate them whether by shotguns or imported viruses, growing in numbers from two dozen in 1859 to 300 million in 1997.[51]

Finally, when ill-effects are recognized, effective remedial action, if it is available at all, will, most likely, be resisted for a considerable time. For instance, it took decades before the potential of chemicals like PCBs for disrupting

---

[50] Martin Teitel and Kimberly Wilson note that "Perhaps the glint of gold on the horizon has blinded the would-be pharmers to the long list of problems that could accompany this kind of technology." Teitel and Wilson 1999, p. 116. For a discussion of profit and social control motives in genetic engineering, see also Hubbard and Wald 1993.

[51] Bryant 1999, Chapter 9.

endocrine systems was officially recognized and a ban incorporated into an international treaty.[52] Much damage has already been done and much more is inevitable since endocrine disrupters are still widespread in the environment and a large portion cannot be recovered. Industry pressure on the U.S. Environmental Protection Agency to retroactively approve StarLink corn for human consumption is an early example of resistance to even acknowledging problems in the arena of genetic engineering.

In view of the magnitude and unpredictability of the risks, biologist Richard Strohman has suggested that "biogenetic engineering of humans and of plants where unanticipated results could cause damage to individuals or to millions of acres of cropland will have to cease except under tightly controlled laboratory conditions and until the time when the complexities are understood and the dangers eliminated. Controls here would include concerns of ethical, legal, and social dimensions. These concerns must reflect the 'ethics of the unknown' of the incompleteness of the science being applied, and not just the ethical concerns growing out of a 'successful' technology."[53] The limitations on inter-species genetic

---

[52] NAS-NRC 1999. According to an international treaty signed in the year 2000, after decades of accumulating evidence of harm, a dozen persistent organic pollutants will be phased out. Jeter 2000. It will take more time to ratify it and more time to actually achieve the phase-out. Of course, for the most part, the dispersed chemicals cannot now be recovered.

[53] Strohman 2000, p. 117.

engineering recommended by Strohman are fully warranted and urgently needed.

When the effects of creating modified genomes on the environment and on evolution (including our own evolution as human beings) are well understood, we can at least have a well-informed debate about genetic engineering. Today, we cannot. We are broadcasting the seeds of possible severe genetic and ecosystem damage without even making a good-faith attempt to know what we do.

We are broadcasting the seeds of possible severe genetic and ecosystem damage without even making a good faith attempt to know what we do.

References

Angier 2000. Natalie Angier, "For Monkeys, a Millipede a Day Keeps Mosquitoes Away," *New York Times*, December 5, 2000, p. D1, D5.

Back from the Brink Campaign and Project for Participatory Democracy 2000. *Short Fuse to Catastrophe: The Case for Taking Nuclear Weapons Off Hair-Trigger Alert*, A Joint Report of the Back from the Brink Campaign and the Project on Participatory Democracy, Washington, D.C., February 2001.

Broad 2001. William J. Broad, "Australians Create a Deadly Mouse Virus," *New York Times*, 23 January 2001, p. A8.

Bryant 1999. Peter J. Bryant, *Biodiversity and Conservation: A Hypertext Book*, Chapter 9. On the Internet at http://darwin.bio.uci.edu/~sustain/bio65/lec09/b65lec09.htm

Campbell 1996. Neil A. Campbell, Biology. Menlo Park, CA: The Benjamin/Cummings Publishing Company, 1996.

Capra 1996. Fritjof Capra, *The Web of Life: A New Scientific Understanding of Living Systems*. New York, Anchor Books, 1996.

Chargaff 1976. Erwin Chargaff, "On the Dangers of Genetic Meddling," *Science*, vol. 192, no. 4243, 4 June 1976, pp. 938 and 940.

Childs 1996. Gwen V. Childs, Ph.D., articles on mitochondria at http://cellbio.utmb.edu/cellbio/mitochondria_1.htm and http://cellbio.utmb.edu/cellbio/mitoch2.htm#DNA

Colborn, Dumanoski, and Myers 1996. Theo Colborn, Dianne Dumanoski, and John Peterson Myers, *Our*

Stolen Future: Are We Threatening Fertility, Intelligence, and Survival?—A Scientific Detective Story . New York, Dutton, 1996.

Darwin 1998 edition. Charles Darwin, *The Origin of Species.* New York: Modern Library, 1998, originally published in 1859.

Davenport et al. 1990. J. Davenport, D. J. Grove, J. Cannon, T.R. Ellis, and R. Stables, "Food Capture, Appetite, Digestion Rate and Efficiency in Hatchling and Juvenile *Crocodylus Porosus*," *Journal of Zoology*, vol. 220, pp. 569-592, 1990.

FoE 2000. "Contaminant Found in Taco Bell Taco Shells, Food Safety Coalition Demands Recall by Taco Bell, Philip Morris," Friends of the Earth Press Release, Washington, D.C., 18 September 2000. On the web at http://www.foe.org/act/getacobellpr.html.

Goldsmith 1998. Edward Goldsmith, *The Way: An Ecological World-View.* Athens, Ga.: University of Georgia Press, 1998.

Gould 2001. Stephen Jay Gould, "Humbled by the Genome's Mysteries," *New York Times*, 19 February 2001

Hubbard and Wald 1993. Ruth Hubbard and Elijah Wald, *Exploding the Gene Myth: How Genetic Information is Produced and Manipulated by Scientists, Physicians, Employers, Insurance Companies, Educators and Law Enforcers.* Boston, Mass.: Beacon Press, 1993.

Jackson et al. 2001. Ronald J. Jackson, Alistair J. Ramsay, Carina D. Christenson, Sandra Beaton, Diana F. Hall, and Ian A. Ramshaw, "Expression of Mouse Interleukin-4 by a Recombinant Ectromelia Virus Suppresses Cytolytic Lymphocyte Responses and Overcomes Genetic Resistance to

Mousepox," *Journal of Virology*, vol. 75, no. 3, February 2001, pp. 1205-1210.

Jeter 2000. Jon Jeter, "Global Ban on 12 Toxic Substances Approved," *Washington Post*, 12 December 2000, p. A42

Kaufman 2001. Marc Kaufman, "U.S. Will Buy Back Corn Seed: Firms to Be Compensated for Batches Mixed with Biotech Variety," *Washington Post*, 8 March 2001, p. A3.

Kaufman 2001a. Marc Kaufman, "Biotech Grain Is in 430 Million Bushels of Corn, Firm Says," *Washington Post*, 18 March 2001, p. A8.

Kimball 2000. John W. Kimball, "Endosymbiosis and the Origin of Eukaryotes," http://www.ultranet.com/~jkimball/BiologyPages/E/Endosymbiosis.html#chloroplast, web page dated 30 October 2000.

Levins and Lewontin 1985. Richard Levins and Richard Lewontin, *The Dialectical Biologist.* Cambridge, Mass.: Harvard University Press, 1985.

Lewontin 2000. Richard Lewontin, *The Triple Helix: Gene, Organism, Environment.* Cambridge, Mass.: Harvard University Press, 2000.

Losey, Rayor, and Carter 1999. John E. Losey, Linda S. Rayor, and Maureen E. Carter, "Transgenic Pollen Harms Monarch Larvae," *Nature*, vol. 399, no. 6733, 20 May 1999, p. 214.

Lovins and Lovins 1999. Amory B. Lovins and L. Hunter Lovins, "A Tale of Two Botanies," Rocky Mountain Institute, 1999. On the web at http://www.rmi.org/sitepages/pid178.asp.

Margulis and Fester 1991. Lynn Margulis and René Fester, eds. *Symbiosis as a Source of Evolutionary*

Innovation: Speciation and Morphogenesis. Cambridge, Mass: MIT Press, 1991.

Millstone, Brunner, and Mayer 1999. Erik Millstone, Eric Brunner, and Sue Mayer, "Beyond 'substantial equivalence'," *Nature*, vol. 401, pp. 525 – 526, 7 October 1999.

NAS-NRC 1999. National Research Council, *Hormonally Active Agents in the Environment.* Washington, DC: National Academy Press, 1999.

Polanyi 1958. Michael Polanyi, *Personal Knowledge: Towards a Post-Critical Philosophy.* Chicago: University of Chicago Press, 1958.

Pritchard 1990. D. J. Pritchard, "The Missing Chapter in Evolution Theory," *Biologist*, vol. 37, no. 5, pp. 149-152.

Sartre 1966 edition. Jean-Paul Sartre, *Being and Nothingness: A Phenomenological Essay on Ontology*, translated by Hazel E. Barnes. New York: Washington Square Press, 1966. First published in French in 1943.

Schrödinger 1967 edition. Erwin Schrödinger, *What is Life?: The Physical Aspect of the Living Cell* with *Mind and Matter* and *Autobiographical Sketches.* Cambridge, England: Cambridge University Press, 1967. First published in 1944.

Spalding and Forrester 1991. Marilyn G. Spalding and Donald J. Forrester, *Effects of Parasitism and Disease on the Nesting Success of Colonial Wading Birds (Ciconiiformes) in Southern Florida*, Final Report, Project NG88-008, Florida Game and Fresh Water Fish Commission, Tallahassee, April 1991.

Strohman 1997. Richard Strohman, "Epigenesis and Complexity: The Coming Kuhnian Revolution in

Biology," *Nature Biotechnology*, vol. 15, 1997, pages 194-200.

Strohman 2000. Richard Strohman, "Upcoming Revolution in Biology," interview published in Casey Walker, ed., *Made, Not Born: The Troubling World of Biotechnology*. San Francisco, California: Sierra Club Press, 2000.

Tautz 2000. Diethard Tautz, "A Genetic Uncertainty Problem," *Trends in Genetics*, vol. 16, no. 11, November 2000, pp. 475-477.

Teitel and Wilson 1999. Martin Teitel and Kimberly A. Wilson, *Genetically Engineered Food: Changing the Nature of Nature*. Rochester, Vermont: Park Street Press, 1999.

Thomas 1974. Lewis Thomas, *Lives of a Cell: Notes of a Biology Watcher*. New York: Viking Press, 1974.

Turner 2000. J. Scott Turner, *The Extended Organism: The Physiology of Animal-Built Structures*. Cambridge, Mass.: Harvard University Press, 2000.

Weinberg 1972. Alvin Weinberg, "The Safety of Nuclear Power," based on a lecture before the Council for the Advancement of Science Writing, Briefing on New Horizons in Science, Boulder, Colorado, 14 November 1972. DOE/TIC—10748.

Wesson 1991. Robert Wesson, *Beyond Natural Selection*. Cambridge, Mass.: MIT Press, 1991.

Wolfenbarger and Phifer 2000. L.L. Wolfenbarger and P. R. Phifer, "The Ecological Risks and Benefits of Genetically Engineered Plants," *Science*, vol. 290, no. 5499, 15 December 2000, pp. 2088-2093.

The Institute for Energy and Environmental Research provides the public and policy-makers with thoughtful, clear, and sound scientific and technical studies. IEER aims to democratize science and promote a healthier environment.

Arjun Makhijani is president of the Institute for Energy and Environmental Research in Takoma Park, Maryland (www.ieer.org). He has written widely on energy, environmental, and security issues. He completed his Ph.D. in 1972 from the Electrical Engineering department at the University of California at Berkeley, where he specialized in plasma physics as applied to controlled thermonuclear fusion.

LiuYu333@Hotmail.com
(sno)

stones22163@.com (henry)

ggxj0122@sina.com (Jeff)

Gao Yong / Xu Jing

Mhaire Ni Mhaolmardhe.